KONG FLU PANDA:
The novel virus that went around the world and the geopolitical tensions that came with it

Copyright @ 2020 He Who Rebels Against All

All rights reserved. No part of this document may be reproduced in any form by any electronic or mechanical devices, including information storage or retrieval systems WITHOUT permission from the publisher, except by reviewers, who may quote passages in the form of a review.

ISBN 978-0-578-69682-9

Fonts by Jess Latham. Thank You.

Printed, Distributed and Bound in the United States of America First Printing May 2020

Published by He Who Rebels Against All

Oklahoma City, Oklahoma 73106

Hey, Thanks for getting a copy!

Facebook.com/TylerLazarus1992

Facebook.com/Oddityler

KONG FLU PANDA:
*The novel virus that went around the world
and the geopolitical tensions that came with it.*

In 2020, a viral pathogen emerges from Wuhan, China.

DECEMBER 31st, 2019 - Chinese Authorities treat dozens
of new pneumonia cases of an <u>unknown</u> viral-genesis.
No evidence exists that this strain rapidly spreads from person
to person.

JANUARY 11, 2020 - A BRAND new decade begins.
Meanwhile, China reports it's very "first" death.

JANUARY 20, 2020 - Other countries and the United States
begin reporting cases with patients in intensive care.

JANUARY 30 - World Health Organization declares an emergency is
happening.

FEBRUARY 13 - 14,000 new cases suddenly appear in the area where it
originated from. Wuhan.

FEBRUARY 21 - Cases rapidly start spiking. The Italian province gets hit hard. 111,000 people become infected.
13,000 people are soon dead.

FEBRUARY 29 - The first reported death of COVID-19 (the name of this pathogen after being genetically sequenced)
happens on American soil

MARCH 3 - Europe begins seeing thousands of new COVID cases.

MARCH 9 - Italy, Population number: 60 million enters full state lockdown.

MARCH 11 - World Health Organization changes the status of COVID-19

from an emergency to a global pandemic.

MARCH 11 - in the United states, Donald Trump bans travel from 26 European countries.

*China has stopped flights from Wuhan, but has now allowed Chinese flights to travel domestically. AKA everywhere else.

Spreading it, Hastening it, Unchecking it.

What the fuck.

MARCH 13 - United States Declares a National Emergency Situation

By March 16 - the country shuts down and enters lockdown
less restrictive in certain areas, protocols of social distancing come into play.

MARCH 23 - NEW YORK CITY becomes a tragic epicenter; population density and close quarters being a major contributor to spread and deaths. People have been flying in and out for two months INTO the city; spreading it, without knowing IT.

By March 26, The United States has 82,404 cases

March 27 - Italy reports 1,000 patients dying a day

MARCH 31 - 1/3 of humanity in inside of lockdown

80% of Americans are behind shelter during periods of the day

APRIL 1 - One Million Cases, 43,252 people dead. 193,000 recoveries. The death rate will soon start spiking even with society trying to flatten the curve.

10,000 Americans die in a month

APRIL 5 - 65,000 people are now dead across the globe from this new virus. Within 4 days, over 20,000 people die.

1,250,000+ are currently infected

April 5 - Morning time: 70,000 dead. 5,000 soon die in a span of a few hours.

April 5 - The first human to animal case is confirmed; A Tiger contracts COVID-19 from a person. The Virus can mutate across species and different immune system pathologies.

It has a kill rate of 3%. The Seasonal Flu has a kill rate of .01

By now, COVID-19 has spread to 183 countries/regions on the map; covering it.

APRIL 6 - 74,000 people are now dead, 300,000 new cases appear in five days.

New York City has lost more people now from COVID-19, then from the 9/11 Attacks.

20,000 new cases crop up on John Hopkins University, in less than 5 hours.

In another five days, 30,000 people die.

APRIL 7 - 80,000 dead. Around 5 thousand die a day now.

Scientists warn that a vaccine cant be created and distributed out for about 9-15 months, depending on speed of processing, understanding COVID-19's weak spots and manufacturing A CURE.

APRIL 9 - 90,000 dead

China, comes under increasing scrutiny for the epicenter. They lied about their figures, saying only 3 thousand died.
It was confirmed that a funeral driver delivered around 3 thousand urns, twice, and estimates now project between 30,000+ dead in China.

That's the very conservative figure.

The World Health Organization's leader covers for China and allowed international flights out of it, while assuring the world it wasn't spreading, after Taiwan warned that it did.

China and Taiwan hate each other; The WHO organization becomes seen as corrupt and a pawn in the pocket of China.

All Credibility of the largest medical agency falls,
The leader himself has no medical license.

Under pressure, in ONE night, China amends it's figures to 4,000; making the lie even more obvious that the numbers suddenly jump over 50%.

Deep political tensions begin, that have been festering for generations between the East, The West, and now the world, all turning towards WUHAN.

21 million cell previous cell phones SUDDENLY go dark from inside of China.
That isn't a death number, but a peek into the level of communistic control they are exerting; Keeping control on their population from letting news or information, NOT curated and controlled by them, likely become The top, internal priority.

Fast Forward >>>>

APRIL 22 - 183,441 humans dead.
2,628,929 cases have spread.

The world is trying to re-open, while navigating through the hurdles COVID-19 is presented. The Entire WORLD economy has shutdown, entered past recession levels. 30 million people are on Welfare, financial markets crippled and staggering. NATIONAL markets. Oil is now in the negative, having no value. Absolute Chaos is Simmering.

"We built"

"A chimeric virus"

With A novel, Zoonotic Spike Protein"

"The spike protein had a dramatic effect"

"To cause disease"

"Seen in human airway cells"

"No current immune therapeutics work against it"

"The lungs are very susceptible to viral Pathogenesis."

"Virus shows in cultures the ability to jump to humans."

"Chinese Horseshoe Bat Populations"

These are the direct bio-containment words used in a Nature research article
that comes from 2015/2016. This is where it ALL begins.

When the virus first broke out, they reported that it Came from a Local Wet Market
in The Wuhan Province.
"Mysterious Corona Virus breaks out from Contaminated Bats in Food Market."

Believable at first, especially with the Known lack of hygienic standards that affect such wet markets with sloppy sanitary conditions.
Not Western prejudice, just plain truth.

Then, as time went on, and the deaths continued to climb,
It was discovered that the ONLY LEVEL FOUR *the highest level of bio-containment*
LAB FACILITY is less THAN 900 FEET AWAY from said Market.

A hop, skip, and not even a jump away.

The Wuhan Insitute of Virology. A center FOR disease control -
Becomes the source of the greatest pathogen release in a hundred years.

"OH no, it wasn't us. It's not the lab. The virus didn't come from there."

Lmao.

Buckle Up. Here's where it gets less Conspiratorial and accusatory
And goes right to Facts. This shit is straight W I L D
Coming at u from da WIld, Wild WEST.

FLIPS OPEN SCIENCE BRIEFCASE

BATS IN CHINA CARRY ALL THE NEW INGREDIENTS TO MAKE A NEW SARS VIRUS - Tina Hesman Saey
November 30th, 2017 - ScienceNews.Org

"We Synthetically Re-derived an infectious full-length recombinant virus

and demonstrate robust viral replication." - Key Labratory of Biosafety, Wuhan Institute
of Virology, Chinese Academy of Sciences. This one is one of the most important ones -
so remember it.

NEW SARS-LIKE VIRUS IS POISED TO INFECT HUMAN BEINGS
The new virus, known as W1V1-COV, directly binds to the same human receptors
as the SARS strain that infected thousands back in 2002.
University of North Carolina, At Chapel Hill, March 14, 2016

throws another pile of papers across the desk for you to read

...to be clear,this virus may never jump to humans, but if it does, W1V1-COV has the
potential to seed a new outbreak with significant consequences for both public health
and the global economy. - March 13, 2016 National Academy of Sciences

This is an illness which attacks respiratory channels in humans/animals
Despite Americans PANIC buying up millions of rolls of toilet paper,
for their asses, This is a pathogen which primarily infects the lungs.
SARS stands for Severe Acute Respiratory Syndrome.
This is SARS-COV-2. There was an original SARS outbreak in 2002/2003
that struck China known as SARS-CoV-1. This is why when it first started
it was an unknown Pneumonia like illness rapidly spreading.
It's a kind of Fatal Pneumonia - at first appearance - with many more malicious
issues that also happen from tissue hypoxia, multiple organ failure, blood
clotting, inflammatory rashes and other conditions that become FATAL
which COVID-19 triggers.

China's ONLY biosafety level 4 Lab is Parked right INSIDE Wuhan.
Where 14,000 people suddenly come up ill.

So What? Where's the proof they engineered this virus or even created it?

IT'S COMING NOW, just wait.

I'm not going to let u intellectually deflect your way out of this, I'm going to keep my knee to my readers neck and they wont be able to breathe. Ironically like what this virus does.

CANNOT BREATHE.

Certain infectious agents and diseases require higher level of bio-containment
safety and lab protocol levels, and some labs cannot EVEN carry certain viral/diseases
inside their facilities unless they receive such clearance for study and intel. The higher the pathogen level, the higher the clearance level.

Because OF the risk for improper handling OR release upon society.

On my birthday, February 22nd, Nature.com publishes an article on The Wuhan
Institute of Virology. It reads. "A Laboratory in Wuhan is on the cusp of being cleared
to work with the world's most dangerous pathogens!"

Followed by photos of Lab Hazmat suits hanging in a closet.

Second Paragraph of Article: "Some Scientists, however, worry about certain pathogens escaping."

EL oh EL.

In 2015, The institute had done research on bat Corona Viruses along with other research facilities which shared their information. By mixing a bat's pathogens with SARS, they found it could infect people by the chimeric build they had formulated from that base.

It was researched and studied.

Not only inside China, but across the world, Bats have been known to be natural reservoirs of disease, meaning they can be infected by pathogens without actually dying to them,
and Chinese Horseshoe bats, in PARTICULAR, act as natural tanks for SARS like corona
pathogens.

CARONA VIRUS.

Batshit crazy.

But WAIT. lol

It gets worse.

These bat C. viruses had been studied in and around the Wuhan Province, right where this sucker emerged from.

"Routinely, they were studied at BSL-2 levels, which is only minimal protection
in Biosafety terms."

Very Dangerous pathogen, incorrectly handled, right near a Food Market.

WHAT could go wrong?????

It'll be fine, stop your worrying, Ethel, and eat your bat soup.

But China - an entire communist country - is not all on the same page.
The South China Institute of Technology determined in April of 2020,
the virus most likely had it's Origins inside the Wuhan Institute of Virology.
Citing very high probability.

But let's just go back up a moment, because one of the most important revelations
of this whole thing will get lost.

Studied and manufactured this badboy inside that lab,
they provided the research notes, they had the extensive body of
biotechnology research to understand what it could do and also its implications
if released into society.

They had the recipe, they just had to bake the corona cake.

April 29th, 2020
217,555 are now dead
3,126,806 infected.

Let's just back up a second

Turn the clocks back to Februrary 5th,
The virus is still relatively unknown,
It's spreading through China but the world isn't really paying it
that much mind.

China, in Ten days, builds an ENTIRE hospital from the ground floor.

645,000 square feet. HUGE ASS BUILDING.
Equipped with 1,000+ beds
7,500 laborers work on this thing.

Looks like a bunch of ants with so many people seen from the Aerial Shots.

It goes up In less than 14 days of time. Finished. Ready TO Go.

A marvelous feat of ingenuity and engineering.

WOW! Look at the fricken feat of it all.

Astounding! Astronomical! That's hella impressive.

Then they build a second one.

(Because The first filled up **that** quickly)
But nobody is thinking of that yet, it's more admiration
Towards China and their capabilities.

THEN Leaked footage - because Lovely Communistic China doesn't allow
freedom of media -
shows less of a large medical facility and ....more of a white, endless
holding room, of bed after bed, after bed, with sick and dying and no
washrooms or bathrooms on the actual
connecting floors.

It's a Damn Death Cell.

No nurses or professionals are seen. Just people dying in a white, clinical bed. Most People just laying still in those beds, no ventilators, no kind of medical paraphernalia, no N95 Masks. It's either people asleep or dead, laid out.

The Death Counter shows each countries numbers AND China's numbers of death don't move or budge for a week. Barely a trickle.
While other countries are having thousands infected and hundreds die every few hours. Italy has 1,000 die in just a night.

China's numbers still don't move. Same Statistic.

It seems relatively confined and safe, to the outside world, That China Has it under control.

China has stopped interior travel, but allowed Domestic outbound flights.

Odd.
Still,
Nobody knows of the hospital rooms and whats inside them.
Just admiration at the pure construction wonder of constructing two, massive hospitals in ten days, each.

Look at Chinese Ingenuity! The footage of whats going on hasn't yet leaked.

Fast Forward to 50,000 DEAD in the United States,

Preliminary drug vaccines are starting with utmost haste.
Commercial use of Remdesivir, created from a chemical Company named Gilead Sciences Inc is used... Remdesivir shows improvement in some cases and no noticeable change in others. Results are inconclusive

- Along with other uses of drugs such as the immuno-supressive

HYDROXYCHLOROQUINE,

[HYDROXYCHLOROQUINE]

A MOUTHFUL TO SAY,
and an anti-malarial drug used for Lupus and Arthritis.
Some doctors continue to use both
As some cases show dramatic improvement.
Practitioners see good results with Hydrox, And Azithromycin;
common antibiotic.

Azithromycin treats a lot of things, it covers things like Pneumonia,
Chlamydia, Cholera, Syphilis, Pink Eye, Strep Throat.

It knocks out LOTS of things, not just sexually transmitted,
but any kind of inflammatory response, it targets well.
It's got a really broad spectrum of uses, paired with the Hydrox.

So those two become the best hope, of the moment, while
actual cures and VACCINES are being studied.

Here's where it gets funky, funky, **FUNKY.**

It's leaked that China TRIED to patent Remdesivir,
the day AFTER Beijing officially confirmed that Covid was able to jump between humans

ONE DAY.

Guess **WHO**…. made the application request.

THE WUHAN INSITUTE OF VIROLOGY.

:
:
:
:
:
….

Can you say cover up, times sixteen and a half.

Oh, but just wait.

China first starts silencing doctors who dissent when they record new cases of a Pneumonia outbreak, A novel CoronaVirus is issued a warning to the medical
community from Dr. Li Wenliang. Chinese Authorities went after him for questioning.

This doctor tries to sound the warning bell and is immediately silenced. He was professionally reprimanded for telling the truth. He was forced to sign a written document
essentially saying that his medical truth was a false rumor he started.

TRAGICALLY,
Dr. Li Wenliang, ends up contracting the virus, after TRYING to make a

plea <u>and dies</u>.

He posts a photo from his hospital bed with a breathing mask, eyes showing
the final looks of a medical extraordinaire who went above and beyond
to TRY and do something.

He was 34 and expecting a second child at home with his wife.

The horror and hypocrisy are only JUST starting.

In another leak of video footage, Doctors are seen crying on their lunch "break,"
and saying in Chinese that there are too many people sick and they themselves
are now waiting to die because of how quickly it spreads from person to person; with not enough safety gear, they all worry and think they will soon fall ill to it.

Genius Medical professionals giving it their all, likely now dead,
after that video was seen.

They're probably all gone now.

China, disgustingly, and not surprisingly after silencing Li and leaving him to his fate,
awards him the "honor" of Martyr and say that he sacrificed himself for the prosperity and betterment of their nation.

Beyond words to describe the lack of ethical truth running in that official government
"award."

Li tried to raise the warning BACK in DECEMBER of 2019.

Thousands of Chinese citizens react against the government in Rage across chinese
apps and social media pages, but again, because China has full digital control,
most of it becomes immediately suppressed and censored.

Deleted.
Gone.

Not a trace, because that's what China does best,
unless it complies with their narrative of how they want to be seen.

+++

A Woman living at the epicenter of all this, Fang Fang, begins
recording daily diary entries and she sees people die around her in homes and apartments.
Each time, The Chinese Party begins blocking her pages from public view and access.
She's already gotten so many views, they can't keep up with the level required to stop her.

YOU GO GIRL

She persists and goes onto complete a diary as a central figure inside this viral
outbreak; securing her work into a book that receives massive criticism from China
and likely putting her life in direct danger; it's awaiting to hit shelves.

++

Video footage shows people being dragged out of their houses and apartments in Wuhan

by People in protective Lab Coats, other videos show houses and doors being welded COMPLETELY shut.
Were they creating a central access point to control traffic,
or sealing people to die inside their residences?
Because it looks like some people were sealed inside their tombs.

It gets so much darker. And Darker.

Fang bin casually records eight body bags inside a car,
A truck driver admits delivering 2,500 funeral urns in one shipment to a Chinese News agency.
He then makes a following shipment again.
That's 5,000 dead, right there.

That's one day of "deliveries."

Oh, but as all this is happening, China's death count STILL doesn't move.
The official word - from them - is about 3,000.
The figure mysteriously doesn't alter for nearly two weeks,
and then suddenly gets amended, two months later to a very quiet 4,000.

The likely death toll inside the country at the rate of funeral delivery
is estimated between 15,000 - 60,000 - 100,000 dead.

Man being filmed drops dead in the street, onlookers scurry away in face masks.
Body wrapped in plastic cooking wrap carried out to a dumpster by Hazmat suits.

21 million cell phone users phones go dark during this period.

It doesn't mean 21 million dead, it means the government is exerting massive
levels of control as CHinese identity is directly tied into their cell phones.

What is managing to leak is small snippets, and recorded audio, brief moments of footage, only showing the horror of little pieces, and then it becomes quickly deleted. Forever.

China is watching every piece of media that comes out, like a hawk, but of course, its not humanly possible to monitor EVERYTHING.

So some things manage to escape in terms of news and intelligence.

Otherwise,

No Word In, No Word Out.

But let's take a look at what's happening outside Wuhan, China During this time.

In Italy, 1,000 people are dying in 24 hours, for days at a time.
27,359 Italian people are now dead, 202,000 infected.

Lombardy region gets hit the worst with 13,575 dead soon after.

Spain loses 24,275 people.

Madrid has 8,048 perish.

United Kingdom loses 21,678. France has 23,660 people die.
The United States has 58,964 people perish.

New York City loses 17,000 civilians at the time of this writing.

New Jersey toll 6,442.

These numbers will be higher by the time you read this,
because people are still dying in THESE areas.

Everywhere you go, it starts taking names right + left.
At first, it was believed with a fatality rate of 3% to only kill
immunocompromised individuals and those with preexisting conditions
but the way that the virus interacts with the immunity varies from person to
person.

People with no prior conditions dropped dead, healthy people in their early
20's
and 30's couldn't beat it. Went down like that *snaps fingers.*

Nations operating under their own sovereignty begin to push plans to
reopen their
countries and struggling economies despite the threat not being nullified.

The realistic truth of it all is that people weren't waiting for the virus to
suddenly disappear, but for the hospitals to not become too overburdened
to tackle the massive wave of infected.

That was THE main fact.

We would attempt to flatten the curve - to enforce social distancing to limit or hopefully curb the number infected - but with no known cure possible, and an anti-virus not ready for trial for 12-15 months, it was the best we could do.

With 200,000 now dead+ millions would have died had social distancing measures,
masks, PPE gear and other measures not come into play.

3 million infected, we currently expect the number of deaths to steadily increase
as testing for COVID becomes more prevalent.

From an immunology perspective, when we talk about people suddenly dropping dead or flatlining in the streets or in a hospital room, it does an injustice to the broad and complex series of events which are taking place, internally, inside them.

The specific rationality of how COVID-19 impacts the body is the way in which it causes
a massive storm of inflammation inside the lungs, connecting tissues and other organs.

The now known 'CYTO STORM' which has been much discussed explains it in this complex airway and human lung cell vivo mucous movement

But again. You have a virus with specific architecture SIMILAR to HIV, which also attacks the entire body, systemically. It -Sars-Cov2- is a full body experience,
while taking a sniper headshot, kill-shot to the human lungs

"With specific damage done to airway tissue." -- From the science notes that created it.

NEW SYMPTOMS appear as the weeks go by, and as knowledge is increasing about
whats happening.

Because you have COMPLICATED Viral Architecture similar to HIV and other lethal viruses,
It's the reason why no known "cure" can magically just pop up.

You have a… pathogen ….
jumping beyond six feet of distance, from person to person, from their own air particles,
and able to spread from smog and environmental pollution as particulate matter
just ironically going after VERY specific Lung cells to target the hosts body to
create a shutdown and die;
Viral traces Lingering on stainless steel, cardboard, clothes and other materials
for hours and DAYS on end.

I'm gonna get to the more specific side of this virus on a cellular level, in a bit, but we know that it WAS
Designed originally from the remnants of the first SARS outbreak, and bonded
with a Bat's reservoirs molecular backbone for added disease,
and twisted under genetic sequencing into a brand new monster.

An absolute disaster was waiting to unfold inside that little, glass vial

labeled with a red warning sticker.

Those people screaming in agony for help in those Wuhan Hospital Beds,
Some gasping for air, others still and silent, suffering in their final hours
with no loved ones, medical staff or anyone there to comfort them.

Oh, And china starts blaming and IMMEDIATELY deflecting TO other countries for "creating" it, insisting
they brought it over there. That CHINA is the victim, after putting their own citizens through absolute

 PANDA-emonium sealing people inside their homes

against
their own will, dragging people down the road in body bags, and throwing deceased into a dumpster.

It reminds me of a man who cheats on you - gives you an STD - and then BLAMES you for getting it, and accuses you of being the ONE sleeping around…. after all the injustice of it….

BUT

Nevermind the fact that their central viral institute started silently seeking patents
for a potential cure, and that they had previous Corona viruses Inside this building
that they HAD been working on stored with 96%
similarity to the new, created form.

Hilarious-ly sick.

All it needed was to mutate inside another host and take off,
to ensure that final 4% terrible percentage.

A carrier.

Patient Zero.

3,452,285 Infected
A Quarter of a Million People Dead +(Not including China's hidden Tens of thousands)
70,000 Americans Now Dead

Ladies, and gentlemen,
The president of the United States

"... And then came a Plague, a great and powerful Plague, and the World was never the same agan! But America rose from this death and destruction,

always remembering its many lost souls, and then lost souls all over the World,
and became greater than ever before."

I don't believe in The Christian God, who set up a Pyrmaid scheme,
But GOD Help us.

October 25, 2019 Before THIS all started - New York Times Article

IN a move that worries many public health experts, the federal government is quietly
shutting down a surveillance program for dangerous animal viruses that someday
may infect humans."

Ironically written By another Donald: Donald G. McNeil Jr.

I'm not going to be a critic of Trumps presidency, that is a job for the historians
to piece together among all the polarized and divided pieces that America has now become, but what I will say is a strong leader without the bullshit is needed
and isn't currently there.

Where Trump did succeed was shutting down Travel from China, that China was grossly trying to push. He was called racist, but likely saved Thousands of Lives From implementing it. Where he failed was not taking this seriously enough,
playing it off as a democratic hoax, and telling people to drink bleach as a cure.

Second part of Donald Mcneils article says "Ending the program, experts fear,
will leave the world more vulnerable to lethal pathogens like Ebola and MERS

that emerge from unexpected places, such as bat-filled trees,etc.."

BAT BAT BAT
BATSHIT CRAYZ
But we got not **Batman**. We got an orange man.

In other news, in the news, "Wuhan Biosafety expert admits widespread security
and maintenance concerns at China's Top-Secret Labs which carry out research
on lethal diseases."

-- Glen Owen, For the Daily Mail 2 May 2020

A dossier comes out which shows China made the initial whistleblowers ..... disappear.

They're probably dead by now, either from COVID or China's supreme love of citizens speaking to truth

As foolish and Childlike as Trump is, He's not the Villain here.

It's China, whose either carelessness or sly maliciousness has now impacted 185 nations.

Throwing such heavy words comes with natural geopolitical tensions in it, but what else can u really say about this whole situation?

It was BOTH a screw up AND a fuck up.

For the record, 2,909 people died in America last night because of their mess
they literally created and engineered in those labs and those science papers
While Trump is tweeting about the "lost souls."

Western Bias or Truth, regardless of which fricken plot of land you inhabit.

What the Fuck

US, The UK, Canada, Aus and New Zealand intelligence agencies have exposed
a series of coverups since this was written from China

Evidence was destroyed, a employee that is believed to be patient Zero,
her profile was removed from the Wuhan Site, but archived by the web,
But China says she moved and is currently in good health,
but that nobody may see her.

LMAO

SHE GONE

…..HOMEGIRL HAS LEFT THE BUILDING -
OUT OF THIS UNIVERSE
AND OUT OF HER BODY

SHE DEADDDDDDDDDDDDDDDDDD.

DEAD.

The bombshell dossier shows China refused to hand over samples to create a cure,
and censored news of the spread and the virus itself since it Started in December
THEN STARTED BUYING UP ALL THE PPE (protective gear)
WHILE SUPPRESSING THE NEWS FOR WEEKS BRO

AND THE WHO (World Health Organization) IS COVERING

FOR THEM - AS THEY ARE - DOING THIS.

THE WHO…

--
WHO THE HELL DO THEY THINK THEY ARE

The world is now five months in, in a collective HELL
we're predicting a revenge spike in the fall + winter,
because now that this thing is unleashed it will have seasonal dynamics
just like the flu or any other illness.

That's the thing, this isn't just a one and done kind of fuckery happening,
Live through a wave and that's the end, this fucking thing is here to bleepin STAY

Let's just give you a quick rundown of JUST how fasT this shit has moved

January 25 2020 - Total Deaths, 22
February 16 2020 - Total Deaths, 1,473
March 3 2020 - Total Deaths, 2935
March 10 2020 - Total Deaths, 10,711
April 1 2020 - Total Deaths, 62,050
April 7 2020 - Total Deaths, 94,563
April 18 2020 -Total Deaths, 154,217
April 25 2020 - Total Deaths, 196,904
April 28 2020 - Total Deaths, 213,810
May 3 2020 - Total Deaths, 242,995
May 11 2020 - Total Deaths, 285,971
May 14 2020 - 4,387,438 Infected.

When we talk about rising tensions within nations,
directed at other nations,
and subtle micro-agressions,
It's best to take a step back at the bigger picture and look at the little details moving those geographic gears

Let's take a look at some of the key players in this whole thing

- Trump - American President

- Dr. Anthony Fauci - CDC Disease Specialist

- Xi Jingping - President of Communist China

- *[Wuhan Institue of Virology]* - *Center Stage*

- Tedros - World Health Organization Leader

Even now, as this book is being written, the world is undergoing revisionist history
as to the origins of the virus. They are covering up a cover up.

I'm about to flip this whole dirty ass entire bed over.

Let's talk some basic Immune Pathology.

Corono VIRUS enters through the host cells by a specific protein spike that covers the outer surface, if you look at photos of it and depictions, it's got these
raised, Edges across the entire cell. That's your protein spike, baby

At the end of each of the spikes, the amino acids of the virus are able to bond

and bind with a host body and those spikes open up specific pathways inside the lungs
Like someone Lockpicking their way through a door.

The first SARS mess did the same exact thing, but in SARS-CoV-2, scientific experiments with the backbone of the virus, messed around with it to create the strongest lock pick possible.

They enhanced the tip.

The Protein Spike.

This includes a sequence of TWELVE Nucleotides that the other SARS and Corona Viruses do NOT possess to them; it's a viral alteration.

Those 12 nucleotides are what give this little fucker the PUNCH it has.

During this writing, China is slamming back against US and world agents saying
They are "evil" for suggesting the Virus came from a lab, but are oddly silent
To their own silencing of whistleblowers and making citizens who raised an alarm
disappear, as "not evil."

Funny little (big) country, they are.

So this new pathogen just happens to have a specific tool to penetrate into lungcells, written in the science papers about its power to target human airways,created and tinkered through lab notes, but we still wanna pretend this just HAPPENED to appear, miraculously, inside nature.

Just popped up.
From one lil ole' bat hanging upside down.

Nobody doubts it came From bats, the genetic sequences DOES NOT lie, There isn't even any room to argue about it, but having a previous Corona Virus
Manipulated and Turned into a Chimeric Beast, with those exact words USED
seems impossible?

Lemme put it this way, Once the cake was baked, all anyone serious scientist had to do
after was read the instructions and re-make it.

Those Notes are the pathway to the Pathology.

Here's where the players start to get Interesting

Dr Fauci says theres no evidence that the virus came from Chinese Lab, but…. admits it came from China

And the chinese lab in the entire country that this happened to be studied out, becomes
omitted from his statements

LMAO. Oh MAN.
OH MAN OH MAN OH MAN

THESE MUGGAFUCKKKKKKKKKKKKAS ARE LYING THEIR WHITE ASSES OFF

Now here is where this guy gets even shadier

He goes on RECORD in 2017 and literally SAYS - SAYS THIS -

"Donald J. Trump will be confronted with a surprise infectious disease DURING his presidency."

BITCH, TIMES FIFTEEN, IN ALL CAPITAL LETTERS

WHAT.

"We will definitely be surprised in the next few years." He Says

BRO, he says this on VIDEO
Not just a misconstrued statement in a paper,
The FOOL goes on camera and says this ISH

"A SURPRISE OUTBREAK"

Oh God, But WAIT.

Let's have a quick review, shall we lmao

283,861 dead
4,142,870 infected
14,531 people passed while editing TWO pages

YO, WHAT THE FUCK LMAO

It's (the rate) drastically slowed Killing - on the first wave -
because the entire world has been basically hiding, but they are worried
about the second wave since everyone is restless and wanting to get the H out
of their house,
uncaring about social protocols to stay safe.

People In Oklahoma, Stillwater
Threatened violence if forced to wear a mask,
then forcing legislators to turn back the rule,
AND then again, one day later, In Oklahomo,
A woman shot two people at Mcdonalds… This Bitch….
for being told she couldn't come inside the dining wing.

Yes, people are that stupid.

Oklahoma, where the COVID air comes sweeping down the plain.

China Lied, People Died,
But here's the update on whats been happening with the virus, itself.

1. Cellphone data in high-security area of the Lab goes dark,
showing no activity in the area where the leak is believed to have happened,
Has a major drop off of people making calls here in October of 2019.

This comes from a report of intelligence agencies and analysts having access
to the public data and reviewing it,
and seeing a mysterious gap in people using data inside the area.

BIG, SMOKING GUN.

The WHO continues to lie, saying it came from the market, 900 feet away, spontaneously,
even though the bats that cause THIS strain arent/werent sold IN the market
at the time of year it started.

SHOTS FIRED

Patient one, or Patient Zero at the BSL-4 Lab is still thought to be the female

who went missing, Author and chinese expert drops bombshell evidence Against Chinese statements and the World Health Organization showing it couldn't have
come from the infamous wet markets

Professor Clive Hamilton is the author of the paper saying the idea that it started In a food stall is quite literally batshit, because the very first people who were infected didn't EVEN have contact WITH the Wetmarkets.

So the plausible explanation is someone got infected inside the lab, ventured out, whether they knew they were infected OR not,
and BAM - rapid transmission from person to person (to person to person)

And millions of people infected later, and over a quarter of a million dead,
The medical community is believing this much more than CHHHY-NUH
and The Flop Health Organization lying their labcoat asses off.

Skepticism of both country and organizations mentioned is AT an all time high,
let's just say that.

Some wild stories I'm watching while writing this

Dude says - That advice to take this seriously probably saved his dads life,
Workers at the company began to get sick and show symptoms,
Couldnt get testing, management at the company didn't take it seriously,
the director ends up in the hospital and dies, soon after, while he had
been coming into work but was trying to hide his symptoms,
Was an epidemiologist and was thought to be a member of the W.H.O.

We've got pathologists and Biotech specialists confirming what seems to be
a Nefarious nature to the nucleotide sequencing they are seeing of the
RNA,
meaning the more dangerous aspects of it showing a genetic insertion
rather than a natural mutation of viral loads.

The evidence is becoming less hazy and more pointed that it came from a lab,
in every form,

BUT SOMEHOW, it came from China but not the Lab in China.

L O L. The mental gymnastics are absurd.

Oh And the Doctor Who is in charge, backed the Wuhan Lab
With millions of dollars
for Corona Virus Research, The same Peckerhead who says a Surprise outbreak will
happen ------------------- ON VIDEO.

BITCHES, how god damn dumb do yall have to be????

How far in the sand, and up your own ass do you HAVE to be to not connect
the MOTHA FUCKIN (DIP AND) DOTS

YA, GOD DAMN DUMMIES

Im going to THROW UP

The National Institute of Health (NIH) with the backing
of the National Institute of Allergy and Infectious Disease (NIAID)
LED BY FRANKENSTEIN FAUCI,
gave THREE .

POINT.

SEVEN.
MILLION.

DOLLARS for bat corona virus research.

$$$$$$$$$$$$$$$$$

The work focused on gains of function,
If you dont know what that TERM is,
It means manipulating pre-existing viruses
to see if they can infect PEOPLE

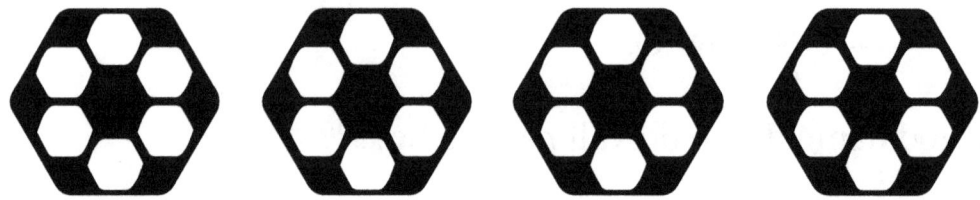

MY GOD.

MY FUCKING. GOD.

    The work got shelved by Obama with a Moratorium TO STOP
because it was seen to be **too dangerous.**

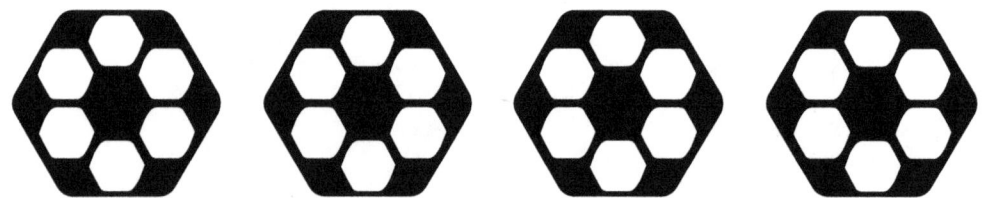

Of course the doctor is saying the virus didnt come from there,

HOMESKILLET threw the funds at the lab for them to even be able to DO IT

Let me just tell you something,
Bless the American people but they CAN be clueless as shit.

Right now, they're all over Youtube and Facebook,
Posting their theories of whats happening,
Flip Flopping from one thing to the next,
And the Goddamn Media isnt doing any better

A stolen plot of land full of thick heads, all yelling at each other.

Fauci has been an adviser to every UNited States President SINCE RONALD REGAN, Was in the middle of the HIV/AIDS mess in the 1980's/1990s

This virus begins in the exact spot that the Lead Medical Doctor is funneling
millions of dollars, to study the ways in can better jump to humans,
THE FOX IS GUARDING THE HEN HOUSE

The leading doctor of the US CORONA TASK force,
Has now entered quarantine himself, because of staffers testing positive for it,
Close to Trump.

Two years before this shit went down, US officials visited the Wuhan lab and
sent back messages - on PUBLIC RECORD - asking for help about improper safety protocols happening
with the risky studies happening inside.

2018 Motha FUCKAS

FROM the WASHINGTON POST

"The cables argued that the UNITED STATES should give the Wuhan Lab further support,
mainly because its research on bat Coronaviruses was important but dangerous."

DING DING MOTHA FUCKIN DING,
YA DAMN DING BATS

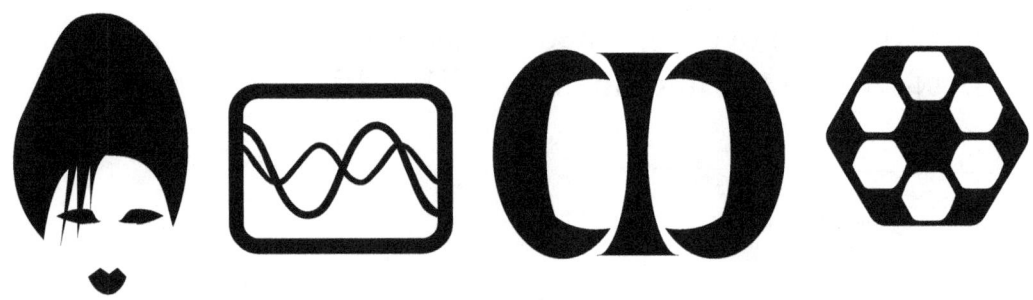

At first, the consensus was that it wasn't a bioweapon, It's Not,
but the protein spike creations show signs of... engineering,
or at the very least, "Creative Tinkering."

None of the relatives to SARS have the same wiring, making Sars COVID, unique
from within the furin cleavage sites, which is outside my paygrade to explain
and probably outside your intellectual standpoint to process.

Came from China, in a Chinese Lab,

With a proverbial MADE IN CHINA stamp on it.

But let's take a step back from a nationalistic charge,
and take a look at the three males who made this situation about five times worse

Tedros - Tedros Adhanom Ghebreyesus - becomes an accomplice to Chinese ...
Maliciousness by being in China's backpocket, while they are supposed and INTENDED to represent the world's health issues and not just ONE countries.

This fool is all over twitter tweeting about "Peace, Love and Unity,"
While Thousands are dying because of his coverup and inaction to speak the truth.

A sick Mo-FOH.

Taiwanese doctors began raising the alarm, the WHO ignored them because
Taiwan isn't represented in the WHO and because China HATTTTTES them.
Whatever, that's between them and not something I wanna get into.
You can spend an entire book just discussing why/how long those two have been at it.

BUT -
There are so many geo-political things/tensions happening all at once, which has only made the initial situation confounded by extremes.

The pandemic had already spread far and wide, invisibly, before the WHO, led by Tedros finally steps up and says on the world stage -- "Oh, shit."

It's easy to criticize someone for their actions when you aren't in the heat of the moment, HOWEVER, to echo the words of China on a world

platform
as your own, assuring the world that everything IS all fine - not because you think it is -
But because China is trying to pass OFF a lie, and then passing that lie for them

That's corruption, times sixty six.

There are photos all over the place of Tedros beaming with a grin from ear to ear, smiling with the Chinese "President."

Lmao

Ok, Now onto Trump - Even Bigger LMAO

What do I even say here? What is historically relevant to his legacy as Twitter Brawler, Game show host and American president???

People are dying and he's off tweeting about low TV ratings of those who dont like him.

Any chance for this man to be petty, he will take.
Any chance.

Is there a chance of him being reelected? Some say YES, some say no chance.

He's on twitter, right now, as Im writing this saying some shit about how amazing of a president he is

Lmao.

I can't.

Then you have Fauci, trying to keep the economy shut down longer, while the country is in the death knell of financial ruin,
The same guy who funded and funneled millions to the Wuhan Lab where this bitch got studied, released and spread.

Major airlines are losing something to the tune of 734 million dollars a day.
??????????????????????????????????

Jesus Fuckin Christ.

Two Alpha Males, One Beta Male
And a Shitstorm of absolute fuckery happening

Russia is getting the piss knocked out of them, the virus took longer to take off over there, Kremlin's chief spokesperson is hospitalized,
Fauci is in quarantine with other Trump Staffers,
New Cases return back to Wuhan after it drops down to Zero,
Resurgence of those with antibodies suddenly becoming sick again,
Raising the question of how long does Immunity last if it returns to the same people who fell ill and then survived and THEN got it again.

We're in a right fucking mess of things.

Every job has masks, sanitization stations, temperature checks at the door to permit or tell someone to leave the premises if they fail it.

Tedros, Trump, Fauci.

One lying, One praising himself, One doing shady shit in labs with shady connections

This cannot be the help we have in THIS kind of a situation.

ARE u fuckinG kidding me.

WHERE OUR ARE LEADERS

Potential Viral exposure, Reopening measures, Safety Protocols,
are the things happening under the underbelly while those 3 tools continue
off being global tools.

Tweeting, Talking, Televising Tools.

Do I think they're idiots, Yes.
Do I think they're foolish, Yes.
Do I think they're evil? Only maybe one of them, truly is.

Behind all the political grandstanding, bloating their own egos,
I see regular guys climbing for power and trying to be authoritative.
Two of them simply got in over their head, the third knew exactly
what they were dealing with - and daresay - got so excited they nearly
wet themself, by saying a surprise outbreak would occur, on CAMERA.

But only of them rings a really sick chord with me,
beyond the normal slimy politico-garb.

But there really is a takeaway besides a global push + pull happening,
or even Bill and Melinda gates, and all these other figures,
with possible darker motives moving pieces,
It's about the fact that nothing can be trusted.

Not the media, not the politicians, not the people.

It's sad that this kind of warranted paranoia has come down,
Because when you compare notes and watch this all in action,
Everyone's acting fuckin shady. Smiling with a knife behind their backs.

The virologists look guilty, the media looks guilty,
these high ranking doctors dont look innocent,
It doesn't look good from any angle

It's like a kid with chocolate all over their face,
and asking them "Did you eat the cake???"

And they immediately say "NO!"

Whose fooling Who here????

Their word isnt worth shit
Tedros saying Its all safe, then saying.. Just Kidding
Trump saying it's a hoax but then BOLDLY proclaiming itll be gone by April
Fauci saying its not from any chinese lab, but funneling cash to the BSL4 Wuhan lab.

It just...... this is about as bad as bad can get.

Never in my life EVER have I sat with the conspiracy theorists, the right wing
or left wing, Ive found them all but especially conspiracy theorists
to have poor argumentation, a lack of higher critical thinking,

the dots they connect just fall to pieces.... But, the truth is,

I don't trust a single one of these rooster--lickers.

This is chaotic now,
but it has the seeds to get EXTREMELY ugly later.

Wars have been fought over less, and I hope it doesn't come to that.
But this is some pure global fuckery happening,
Nobody capable is in charge,
Everyone is lying,
Everything is double-sided and duplicitous
and the everyday man and woman are out here suffering,
Small businesses are closing for good,
They cant weather this kind of a storm out
Bleeding cash, as these corporations make grabs for corporate welfare,
Screwing everyone else over,
And an economy in a capitalistic rut, with numbers not seen since the great depression of the 1930s.

I mean this is a prelude to something BAD.

The big thing is, behind all these small pieces,
Someone knows.

Someone knows how this started, How it began, how it was created, how it was leaked.

Someone knew, if they aren't already dead.

Another body rotting to add to the hundreds of thousands.

It was somebody,
Who do I trust, Nobody,
Repeat after me,

THOSE LYING FUCKS,
now sing it - Everybody.

Any questions,
Anybody? VIRAL ANTIBODIES

THOSE LYING SUMBITCHES

country guitar twangs

THOSE LYIN SUMBITCHES,
COULDNT TELL THE TRUTH
IF THEY TRIIIIIIIIIIIIIIIIIIIIIIED

tumbleweed of lies blows across Asia and America

I don't trust any---ting that comes out of anyone's mouth, anymore.

I think it's smart to verify everything.
I don't think the luxury of believing in the good can exist
when there's so much bad at the higher levels.

I've seen more goodness in regular people, helping each other,
then I have ever seen in the notable ones.

I think its the whole power corrupts thing,
When someone doesnt have anything to gain, their motives are different.

The duality of mankind is horrifying.

I think it's this civilization that glorifies the light,
that so MUCH darkness exists.

People are GOING to feel, that I let Trump off, easily
while Grilling Fauci over the third degree,
Here's the thing, okay.

Trump is a HOT, damn mess.
Four years into his term, I expect nothing less
than the messiness he has displayed. It's not even shocking anymore.

It's routine.
It's who he his.

Fauci, who has thirty years and several administrations worth
of CDC service to his country,
I'm not gonna denigrate from his work, or his service,
those things are beyond any kind of criticism.

What I am critical of, is the most respected health official - in the world -
Making jarring statements about surprise pandemics.

It's like he was so excited, in the moment, he told on himself.

I think he'll try and explain what he "meant"
retroactively, but the pathogen cat is out of the bag.

I don't trust someone who would proclaim that a president is
DEFINITIVELY going to have to deal with a SURPRISE pandemic during
their term.

That's fucked up.

The thing is, Trump isn't smart enough to register he may have people around him
with THOSE kind of intentions.

And when you have a doctor who has been vetted through multiple

presidents,
and WORKED on the HIV crisis, you naturally believe he's benign and good.

But saying what Fauci said, is not good.

And his excuses arent good enough to explain it.

On one hand, he can go "Oh, no people may die, I tried to warn you - you should
have listened,"

And on the surface, it seems like he is chiding people,
but deeper, it seems like

"Yeah people are going to die and I knew it before it happened,
because I had my hand in it, on some level."

Now the big thing is a deflection of saying well the money sent to the Wuhan Lab
Was part of a grant that other sub grants,
but the money trail, no matter how you wanna slice it up,
It doesn't lie.

It all points to this one place.

And of course, funding is going to be sent to laboratories for studies, that's what
THEY do, thats the very purpose of grants and funding,
However,
When these specific Bat Caronaviruses were given funding,
you can't turn around and say,
A Bat Carona virus did not develop here.

Like ?????????????????

You can't be that dumb, you just can't.

And even if you are THAT dumb,
I'm not gonna let you.

An infectious disease specialist has to specialize in understanding the mechanics
of the thing he studies, just like a car mechanic has to understand how cars work
to fix them.

But what if that knowledge was perverted and warped?

Instead of understanding to learn,
it became understanding to damage, to harm,
to kill.

It's all so Godamn Shadowy.
And I'm not saying they did or they didn't, or he said, she said.

I'm just saying it is HELLA troubling.

And hella fucked up.

American Money to monkey around with a new virus where bats were already being
experimented on, with specific Bat Corona Viruses.

Creation and Spread, matter little,
Because at the end of the fuckin day -
it's out here, and real people are dead.

Airing out these crimes, with a manmade pathogen that spreads through the breath,
Deserves to be aired out in a World Tribunal To decide.

Not just a group of readers,
Not one city,
Its infected and impacted all cities.

The world must hold the world health organization, China, and other people
accountable.

The media, is not exempt.

Everyone wants to claim "persecution" these days.

It's time for Prosecution.

COUNTLESS LIVES.

Economic Hardship.

Improper handling of Deadly Pathogens.

Lies from all sides.

Body after Body in orange bags inside freezers.

Asking China to be truthful here
is like Asking a T-Rex to please eat lettuce.

An Asian city of 11 million,
negatively touches billions of lives.

Lying about their death tolls,
Let me end it on this note.

You SHOULD hate the Chinese regime.
They are so, SO, SO evil.
Communism, incarnate.
And Im not saying Capitalism is some Holy, Saint Angel.

But you shouldn't hate the CHINESE PEOPLE.

The government and the people are not the same.
Those poor people did not get to choose their government.
They have no choice,
which makes the Chinese Communist Party the most responsible,
and evil body
of all the bodies laid to their deaths.

There has always been Western bias against Asian communities,
and covert and over racism,
This isn't a free pass to go and slandering innocent chinese people,
But The Chinese Government
has never been the innocent.

When you have a country, controlling, DIRECTLY, the world health organization,
The other countries represented in that collective have
a RESPONSIBILITY and RIGHT to shield themselves from an emergent epidemic.

This isn't another great depression,
If things don't rebound ---
This will be a Global one.

Oklahoma Bombing - 95
September 11th - 2001
War of Iraq - 2003
American Recession Crash - 2008
Global Financial Offset Collapse, 2020.

(TRUE STORY)
You know, it's so ironic,
When I went to purchase my yearly calendar, that I get before a new year starts,

I picked up my new one and it was called "**Hidden Agenda**." RIP

I thought, my, thats a funny name albeit clever way to make a planner.
I purchased it, anyway.

BUT
When a software designer - Billius Gates, no medical degree
 - can freely speak on COVID-19,
with ties to creating the very vaccines to "heal it,"
but licensed medical professionals WITH THOSE degrees
become censored to speak out.

Bitch, if Billy has no medical degree and can become a leading speaker,
I can too. You hoes.

We have a very big, fucking problem on our hands.
As that same man, Gates, is playing around with cardboard cutouts
in his home of people planning for a pandemic event called "EVENT 201"
In his theory work* - Conveniently, now placed at the top tier to make a
cure. Im not even kidding. It's all a matter of public record.

***All these statements, every paper, every one of these a-holes are all on
record saying every bit of this, writing the words I've shared.****

This information **isn't** even hidden in some obscure online cloud.

It's literally public knowledge.

L m f a o.

I HATE IT **HERE.**
LOL.

Let's get something crystal clear about Fauci, Anthony.
This isn't about pedaling theories of populism, people have
already made misdirected attacks out at him because A. They were angry
they had to stay inside and obey a small DEGREE of lockdown measures -
calling for his resignation - from people considering themselves
armed "patriots" that have no idea or concept of what REAL, FORCED
Compliance actually looks like and how ugly it can get.
And B. Fauci comes under attack again for his work in the 1980s
on the HIV+ Work he did.

Listen.

I don't give a flying fuck about what he did then or did or didn't say then -
We are NOT here to discuss Anti-Vaxx rumor mills and conspiracy/
speculation bullshit. I'm going off what the man said, on record,
HIMSELF.
These people wishing him death threats and harassing him,
had nothing to do with understanding his connections with and TO China
as a virologist; they were just pissed and acting petulant because they
couldn't get out. They are not smart enough to go deeper than that,
And Yes, I'm well aware of their "deep state" ideas,
but these people, in charge, are not and have never been THAT deep.

SECOND of all, saying it's a deflection - for the White House's poor
handling of the crisis - IS a deflection, in itself.

Are we flattening the curve here

or flattening the economy ---- past the point of resuscitation?

Are we just going to try and erase and then WHITEOUT
all the connecting dots, leading back to ONE, central building.

Screw the WHITEHOUSE for a moment, and just THINK, Goddamnit.

A MASSIVE drop of cellphone data happens in October inside the Institute,
monitored by intelligence agencies scanning through public record,
speaking directly to a pathogen breach/release event.

This same institute where it HAPPENS - secretly tries to file a PATENT
on a new drug that is later the very top vaccine seen as a cure -
ONE day after they admit the virus (which they also claim isn't from them)
is spreading from PERSON to PERSON.

You **cannot** just take a random swipe in the dark of the medical world
and say, "Ah, Yes! This medicine will do nicely for something we don't
even know what IT is."

This fucking thing was already out then; they were already sequencing
and studying it for them to START searching for FIXES to IT.

Smoke Smoke Smoke
Fire fire Fire
Look at the next smoke screen
<u>Its a Virus</u>

But its not a Chinese lab virus
<u>But Okay it came from China</u>
But no its NOT from China
<u>Okay it looks like it might have a unique sequence to its origin</u>
Okay but its still not a lab
<u>Ok Wait its obviously out of a lab</u>
But definitely not out of a Chinese Lab
Okay Well Not a lab, but its origins are Chinese.
Wait, lets just say it came out of the Wetmarket...
That's believable!

You think this THING just magically jumped from a bat and appeared in the sky of China one, magic night?????????????????????????
With specific protein spikes all over it that to incredible damage to extremely specific lung cells? That may cause injury that never fully heals. That just appeared one widdle day?????

Get the fuck outta here with that.

How many more redirecting attempts are we gonna allow?
Running in circles, like a sneaky fox, to HIDE the obvious.

Let's go back to the very first words of escape

"In 2020, a viral pathogen emerges from Wuhan, China.

DECEMBER 31st, 2019 - Chinese Authorities treat dozens of new pneumonia cases of an unknown viral-genesis. No evidence exists that this strain rapidly spreads from person to person."

Comes from China, But its not from China?
Right?
Story is said it begins at a Wetmarket, 900 feet away from the highest level lab that exists.
But it's not from a lab? Right?
Not from a wetmarket? But it's... from a wetmarket? Right?
Lab is studying bat corona viruses, with research grants into this specific Corona virus. But it's not THIS lab? Right?
It's a Corona Virus, but it's NOT this Corona Virus????

RIGHT?????

idiots.
fuckin idiots.

The president, the people, the media, every single one of them THINKS they know best.

Lemme break down the hierarchy of this for you, real quick.

It's a Pyramid.

The people at the bottom - all believe themselves experts on everything. Fact is, they aren't. But don't try and tell them **that.**

Next UP is "THE MEDIA" which can barely put together an article online without spelling errors - giving the news of the day.

The president, at the top, is a weird mixture of stubbornness of the people, and the insufferable, never-shuts-up quality of the Media. Inside ONE body.

The media, struggling to keep the shitshow of American news going, somehow sees itself capable of jumping into biotechnology matters.

HAH-HAH!!!!

HAHAHAHAHAHAHAHAHAHAHAHAHAHAHAHA

Lastly, let's UNDERSTAND this. Because 95% of you DONT.

A purposely created VIRUS is NOT THE SAME THING
as a Purposely RELEASED one.

Stop. MY GOD. STOP MIXING THE TWO TOGETHER.

You can't look at those lab notes with THAT wording and see OR say it
wasn't made, but just because it was intentionally MADE,
does NOT mean it was intentionally released.

I had people argue with me that wearing masks were pointless - Even
though China adhered strictly to them to stop the spread - and for the
record, the United States only reversed its official position of saying not to
wear masks, because the general admitted they were trying to conserve
what they could in terms of protective gear for the health officials and
frontline nurses that would be exposed. Once that wave was up, they go on
record, admitting, that masks DO prevent the spread of infected particles.

I had these same people, the tier below the media, have to stop arguing
with me because their states began to ENFORCE the mandatory wearing of
masks, with steep fines, if anyone was caught in public without them;
to prevent the spread of COVID-19 jumping from person to person.

I use this example, because if I don't even trust people that are my friends,
do you think I'm going to trust anyone else?

I love a great number of people, but I trust nobody except myself when I
wake up, simply because, I don't trust people to do the work to verify
what's being said.

Everyone thinks themselves this great, automatic expert - a pandemic of
hubris - and then the actual experts are lying sacks of shit.

The real pandemic going around is the fact that people can pull things from
thin air, reliant upon their feelings and substitute it as an ACTUAL FACT.

You have an entire generation of people considering themselves genius level, or a prodigy of intelligence.

I'm possibly smarter than you, and I am not a genius.
Nowhere even close.
In fact, the people that declare themselves a "genius" are some of the stupidest people I've ever had to have a conversation with.

It's OKAY to be dumb. Jesus, as if THAT's the worse thing to be accused of.

But stupidity, when it becomes collective, and enters the echo chamber of the MEDIA and THE people, and a president which is a hybrid of both….

Good God Gloria Almighty.

Aretha Franklin, Chain of Fools.

Everyone wants some scathing, bitchy comments on Trump, but he's not the target
The thing about Trump is - and I've said this since jump in 2016 when he got elected -
Trump is just a pawn to be manipulated around, He's easily angered, quick

to react,
Tweets his entire strategy to everything in real time. He's not a threat

It's the fact that he's SO easily maneuvered which is why Russia wanted him in place,
Not just Some secret Soviet Ambitions, or Memorandum to destroy Clinton as a threat,
Trump is just easy to play the strings with, because he believes himself nearly infallible to manipulation, is precisely why hes so easily manipulated,
And Likely Why Putin, who works from the shadows, and reveals NOTHING,
wanted a man who does his business in the sun, and airs out EVERYTHING.

Thats why it always amused me when people, early on, said Trump and Putin
were bedfellows. No, Putin is the polar opposite to Trump,
and BTW during the middle of all this shit, Putin initiated a vote
to amend his countries constitution, to REMAIN IN POWER
UNTIL PAST 2036 (The Guy is NEVER going to give up his IRON lock
on Russia. NEVER. EVER. EVER. A Dirty Dictator in Presidential clothing.)

Oh and Btw, on another note of Russian Loveliness, three doctors have "fallen"
out of windows at their hospitals for complaining about the lack of protective
gear to fight this God Forsaken Virus.

Absolute Horror and Hilarity.

Yes, three top-tier professionals who spend a decade in medical school and learning
human physiology, just happen to stumble through glass windows on the top floor

of a hospital, because they're that clueless *EYEROLL to UKRAINE*

Rubs Temples Oh MY God, This whole thing is such a mess

Human Evilness, Human Stupidity, Human Lies
All converging down in one single brilliant flash of time.

NUH-UH, The Coverup of it being created is now reported
NUH-UH, The Chinese Government says about the Virus
NUH-UH, The Americans refusing to wear masks chant

NUH-UH

NUH-UH

NUH-UH

Oh my God, I feel sick

+++++ SUDDEN CUT++++

MARS INTRO

+++++

RETURN TO EARTH

John Lennon's Imagine is playing.

In a fever dream,

The White Man,
The Indian Man,
The Black Man,
The Chinese Man,
The Gay Man,
The Straight Man,
The Christian Man,
The Muslim Man,

Dance in a Large Circle with

The White Woman,
The Indian Woman,
The Black Woman,
The Chinese Woman,
The Lesbian Woman,
The Straight-ish Woman,
The Christian Woman,
The Muslim Woman,

IMAGINE ALL OF THE PEOPLE * is heard *

LIVING FOR TODAY

YOO-HOOOO-OOOOOO

The group of people dance in perfect harmony and peace, for a brief period,
No War, No Anger, No Possessions,
Just love for one another and themselves

They're all butt naked, with wildflowers in their hair
Floppy Dicks and Vah-jay-jay
UNTIL

The Christian Man trips over the Muslim Woman,
HEY WATCH IT BUDDY,
YOU WATCH IT, The Muslim Woman shoves back,
and knocks over the Gay man who ends up getting too close to the straight man,
DONT TOUCH ME, FAGGOT
WHO ARE YOU CALLING A FAGGOT YOU PIECE OF SHIT,
The gay and straight males engage in a fight,
The Dance Of World Peace Breaks up,
The Black Man Shoves the White Woman
UGLY WHITE BITCH,
HEY DONT CALL ME THAT,
THE N WORD GETS DROPPED IN THERE
The White Woman,
Shoves the Chinese Man,
AND
YOU CREATED THIS VIRUS, YOU NASTY CHING CHONG SUMBITCH

All out discord and disharmony breaks out

John Lennon's "Imagine" 𝄞 Becomes Distorted, and Warped

ALL OF THE PEEEEEE
PULL

His voice darkens, distorts backwards

CHERRRRRRING

ALLL THE

John's voice sounds nearly evil in tone

WORLD

The song is no longer heard over threats, screams and shouts

Of the masses, so there are the asses,
White, Gay, Black, Straight, Asian, European, ASSES

People are seen panic buying Toilet Paper on the news

Somewhere a nuke is suddenly dropped.

To the 300,000+ people now killed from this virus in less than 3 months of time.

Rest in Peace

www.ingramcontent.com/pod-product-compliance
Lightning Source LLC
Chambersburg PA
CBHW081349040426
42450CB00015B/3359